I0078891

CSI Historical Bibliography No. 2

LIGHT INFANTRY FORCES

by

Major Scott R. McMichael

Combat Studies Institute

U.·S. Army Command and General Staff College   037260

Fort Leavenworth, Kansas

January 1984

84-2218

CONTENTS

016 .3561
M167L
c.2

PREFACE

This annotated bibliography was initially developed in conjunction with the initiative of the Department of the Army in 1983 to develop the force structure for 10,000-man light infantry divisions. Its goals were to provide annotated historical references for the combat experiences of previous light divisions and to list historical sources on the force design process, especially in regard to attempts to lighten the force or to respond to improvements in technology on the battlefield. The first draft of this bibliography was distributed in September and October 1983 as a quick reference to force planners across the Army.

Since that time, the bibliography has been expanded, but the general focus remains the same, historical light divisions and the force design process. It addresses light infantry forces of the twentieth century with primary emphasis on World War II and later. This document can serve as a starting point for force designers' research into the origins, organizations, capabilities, and combat experiences of light infantry forces.

In order to permit widespread distribution of this bibliography, only unclassified sources have been cited. However, additional classified documents on the subject exist, particularly in regard to technical analyses of force capabilities. A number of these are available in the Combined Arms Research Library (CARL) at Fort Leavenworth, Kansas.

The enclosed charts describe light infantry and close combat forces of World War II as well as some of those in being today. The selection of the contemporary units was based on their capability for strategic mobility, since this is the overriding design parameter for the new U.S. light division. The charts afford an opportunity to discern how different countries have approached similar military requirements, developing military organizations that differed widely in size, structure, equipment, and tactics. Data depicted on the charts has been taken from sources cited in the bibliography.

Since World War II, factors such as mechanization, nuclear weapons, and electronic wizardry have obscured the significant role to be played by confident, agile, fit, motivated, and well-trained infantrymen. Yet, even in this era of high technology, groups of foot soldiers remain powerful and influential forces on the battlefield. This bibliography is intended to contribute toward the rediscovery of the utility of light infantry.

NOTES ON DOCUMENT LOCATIONS

Many of the references in this bibliography are available in large public and university libraries or in military libraries established across the United States. However, a number of them exist only in special libraries or special offices. The author has identified those not widely available by notations and call numbers at the end of the bibliographic citation. The abbreviations below are provided for assistance.

CARL . . . . Combined Arms Research Library
Combined Arms Center, Fort Leavenworth, Kansas

CSI. . . . . Combat Studies Institute
USACGSC, Fort Leavenworth, Kansas

DTIC . . . . Defense Technical Information Center
Cameron Station, Alexandria, Virginia

MHI. . . . . Military History Institute
Carlisle Barracks, Pennsylvania

Beaumont, Roger A. Military Elites. Indianapolis, IN: The Bobbs-Merrill
    Co., 1974.

This work deals with elite military units of the twentieth century.
The author analyzes the reasons for the formation of these units, their
relationships to conventional forces and their parent systems, and their
successes on the battlefield in meeting the goals of their creators. The
U.S. Army Tank Destroyer Corps, the U.S. Army Rangers, the First Special
Service Force, the British Long Range Desert Group, the Afrika Korps, the
Waffen SS, the U.S. Army Special Forces, the Irgun, and national paratroop
units are among the many units discussed by Beaumont. The author maintains
that, given the limited benefits of elite forces to the overall outcome of
major battles and campaigns, their cost has been too high, in terms of the
diversion (from conventional forces) and loss (through death) of the
military's best young leaders.

Binkley, John C., 1st Lt. "A History of U.S. Army Force Structuring."
    Military Review 42 (February 1977):67-82.

Binkley's article addresses the evolution of division organization in
the U.S. Army, focusing on the twentieth century. He discusses the World
War I square division, the World War II triangular division, and the postwar
pentomic, MOMAR, and ROAD divisions. He achieves his objective of
demonstrating how the factors of technology, tactics, and combat experience
have influenced division structure through the years.

Doughty, Robert A., Maj. The Evolution of U.S. Army Tactical Doctrine,
    1946-1976. Leavenworth Paper no. 1. Fort Leavenworth, KS: Combat
    Studies Institute, U.S. Army Command and General Staff College, August
    1979.

Since the Second World War, U.S. Army tactical doctrine has owed its
character to a number of factors, often at odds with each other. Among
these factors, national security policy, new technologies, service and
branch parochialism, and actual battlefield experience were the most
effective arbiters of what the Army's doctrine and force structure would be.

English, John Alan. A Perspective on Infantry. New York: Praeger
    Publishers, 1981. CARL 356.1 E58p.

Small groups of foot soldiers remain even today among the most
influential on the battlefield. In support of this thesis, the author
examines infantry operations and training and compares organization,
equipment, weapons, and tactics of several national infantries since 1864.
Well researched and written in a clear style, A Perspective on Infantry is a
superb history of modern infantries, convincing the reader of the continued
importance of the foot soldier even in this technological age.

Marshall, Samuel Lyman Atwood. Men Against Fire: The Problem of Battle
    Command in Future War. Reprint. Gloucester, MH: Peter Smith, 1978.

Originally printed in 1947, Men Against Fire has become a military classic. Using an eloquent prose distinguished for its clarity of expression, Marshall illuminates the problems of leading men to risk their lives on the battlefield, touching on the physical and psychological aspects of the problems of command in minor tactics. This book identifies some of the most important objects for training programs of light infantry forces and leaders. Men Against Fire uses a number of historical combat actions as examples.

Naisawald, L. Van L. "The U.S. Infantry Division: Changing Concepts in
    Organization, 1900-1939." ORO-S-239. Chevy Chase, MD: Operations
    Research Office, The Johns Hopkins University, 7 March 1952. CARL
    N-16454.573.

The author provides an overview of the changes to the U.S. infantry division from 1900 to 1939. He makes comparisons with European divisions and presents the historical arguments for change, particularly those addressing the needs to maximize mobility and flexibility. He also analyzes the results of field tests and combat experience.

Ney, Virgil. Evolution of the U.S. Army Division, 1939-1968. Combat
    Operations Research Group Memorandum CORG-M-365. Ft. Belvoir, VA:
    Technical Operations, Inc., Combat Operations Research Group for U.S.
    Army Combat Developments Command, January 1969. DTIC and CARL AD697844.

The author traces the origins of the U.S. Army infantry division from World War I to 1968, focusing on key personalities (e.g., Maj. Gen. Leslie McNair) and changing influences on division structure. The triangular division of World War II gave way to the pentomic (ROCID) division of the 1950s only to be followed again by the triangular ROAD structure. Ney briefly discusses the light infantry divisions of World War II.

Ney, Virgil. Evolution of the U.S. Army Infantry Battalion: 1939-1968.
    CORG-M-343. Alexandria, VA: Combat Operations Research Group,
    Technical Operations, 1968.

The modern U.S. Army infantry battalion has its origins in the pre-World War II army. Since then, its organization has been influenced by combat experiences, the advent of nuclear weapons, and major advances in the technology of war. The author notes that the prevailing trend has been toward even higher increases in the infantry capabilities for fire and movement.

Romjue, John L. "A History of Army 86. Vol 2. The Development of the Light Division, the Corps, and Echelons Above Corps, November 1979-December 1980." Fort Monroe, VA: U.S. Army Training and Doctrine Command, June 1982. DTIC and CARL AD-F000004.

This work is the official TRADOC history of the development of Army 86. The section dealing with the development of the High Tech Light Division is especially interesting. It describes the original initiative and guidance for the formation of a light division, the interaction of the service schools, CAC, and TRADOC, the development and rejection of the three original HTLD designs, and the final acceptance, with reservations, of the fourth design (17,700+ soldiers).

U.S. Army Command and General Staff College. Evolution of the Division Span of Control, Equipment, and Tactical Doctrine. [Chart]. Ft. Leavenworth, KS, 9 July 1981.

This source is a chart showing the evolution of the division span of control, equipment, and tactical doctrine in the U.S. Army from 1777 to the ROAD division. Force structures of units and staffs are depicted. The chart also explains how improvements in technical capabilities of weaponry and equipment changed force structures and doctrine.

U.S. Army Infantry School. Infantry in Battle. 2d ed. Washington, DC: The Infantry Journal, 1939. Reprint. Fort Leavenworth, KS: U.S. Army Command and General Staff College, 1982.

Prepared under the direction of Col. George C. Marshall, Infantry in Battle is concerned with tactics of small units, with detailed examples drawn from World War I as illustrations. Its purpose, which it achieves, is to acquaint the reader with the realities of war through the emphasis of a number of important lessons. Some of the issues developed by the authors are simplicity, obscurity, decisions, mobility, control, and reconnaissance. Like Men Against Fire, this book is a valuable source of information for light infantry trainers and planners.

Wilson, John B. "Divisions and Separate Brigades in the Army Lineage Series." Washington, DC: Center of Military History, n.d.

Wilson made a comprehensive study of U.S. Army force structures in the nineteenth and twentieth centuries. His study includes RA, USAR, and NG units and presents the arguments and influences behind organizational changes. The inability of the Army planners to agree with field commanders on the structure for a World War II light division is revealing.

Section II. WORLD WAR II AMERICAN EXPERIENCE

Clinger, Fred, Arthur Johnston, and Vincent Masel. The History of the 71st Infantry Division. N.p.: 71st Infantry Division, 1946. MHI UH 05-71-1946. CARL OVERSIZE 940.5421 C641h.

This source is a yearbook-style history of the 71st Infantry Division during World War II. Of particular interest are pages 11-22 that deal with the division's organization, training, and testing as a light (pack) division in 1942-44.

Diamond, Maynard L., Maj., et al. The 89th Infantry Division, 1942-1945. Washington, DC: Infantry Journal Press, 1947. Reprint. Nashville, TN: Battery Press, 1980. MHI UH 5-89-1980.

Much like the selection above, pages 47-62 of this unit history carry the 89th Infantry Division through its organization and training as a light (truck) division in 1942-43, its subsequent testing in Louisiana and against the 71st Light (Pack) Division at Hunter Liggett in 1944, and its deployment to Europe as a reorganized standard infantry division.

Greenfield, Kent Roberts, Robert R. Palmer, and Bell I. Wiley. The Organization of Ground Combat Troops. U.S. Army in World War II. Washington, DC: Historical Division, U.S. Army, 1947.

Citing original source materials in the form of War Department correspondence (which is not available at MHI or CGSC), the authors describe the efforts made by Maj. Gen. Leslie McNair (CG, Army Ground Forces) and his staff to limit the size of American divisions formed during World War II. They discuss differences between the AGF, the War Department, and field commanders. In particular, the authors describe the U.S. experiment with light divisions in 1943-44.

Luttwak, Edward N., et al. "Historical Analysis and Projection for Army 2000." Pt. 1. Chevy Chase, MD: Edward N. Luttwak, 1982- .

This source is a collection of individual papers, produced under TRADOC contract DABT-58-82-C-0055. Included in this collection are the three papers below, which deal with nonstandard U.S. Army forces in World War II. The papers discuss the organization, training, testing, and eventual use of these forces in combat.

_____. Paper no. 1. "The United States Army of the Second World War: The Light Divisions." 1 March 1983.

_____. Paper no. 2. "The United States Army of the Second World War: The 10th Mountain Division (10th Light Division [Alpine])." 1 March 1983.

_____. Paper no. 3. "The United States Army of the Second World War: The Tank-Destroyer Forces." 10 December 1982.

U.S. Army Ground Forces. "The New Infantry, Armored, and Airborne Divisions." March 1946. CARL N-15338B.

Postwar TO&Es for infantry, armored, and airborne divisions, submitted by the Commander, Army Ground Forces, are described in this document. After approval by the War Department, peacetime strength reductions were implemented in TO&Es. CARL documents N-15338A through N-15338E support this subject with other diagrams and charts.

U.S. Army Infantry Conference, Fort Benning, GA, 1946. "Report of Special Committee on Organization of the Infantry Division." In "Report of Committee on Organization." June 1946. CARL R-13559.

This postwar report of the infantry conference formed to consider the lessons of World War II contains a section discussing special divisions. Transcripts of comments by Omar Bradley, Walter Krueger, Courtney Hodges, James Gavin, C. D. Eddleman, et al., are included.

U.S. Forces, European Theater. General Board. "Organization, Equipment, and Tactical Employment of the Infantry Division." Study no. 15. N.d. CARL R-13032.15.

_____. "Organization, Equipment, and Tactical Employment of the Airborne Division." Study no. 16. N.d. CARL R-13032.16.

_____. "Types of Divisions--Postwar Army." Study no. 17. N.d. CARL R-13032.17.

_____. "Organization, Equipment, and Tactical Employment of the Armored Division." Study no. 48. N.d. CARL N-12875.48.

At the conclusion of World War II, the U.S. Forces European Theater Headquarters convened a General Board, composed of some of the most prestigious general officers and commanders from the war. They met to record the most important lessons of the conflict, producing numerous separate studies as a result. The four studies above contain their recommendations for the organization, equipment, and tactical employment of the infantry, airborne, armored, and special divisions.

Several 10th Mountain Division unit histories are on file at Military History Institute, Carlisle Barracks, PA.

Devlin, Gerard M. Paratrooper. The Saga of U.S. Army and Marine Parachute
     and Glider Combat Troops During World War II. New York: St. Martin's
     Press, 1979.

     In 650 pages, Paratrooper presents a detailed account of every airborne
and airland operation conducted during World War II by the U.S. Army. The
book combines extensive research, based on thousands of interviews, with a
stirring narrative. However, its descriptions of airborne operations suffer
from a lack of maps, and little attention is given to airborne tactics or
force structure.

Edwards, Roger. German Airborne Troops, 1936-45. Garden City, NY:
     Doubleday & Co., 1974.

     Unlike some popular accounts of airborne operations in World War II,
this book deals comprehensively and systematically with the uniforms,
weapons, equipment, training, organization, air transport, combat
operations, unit histories, and key personalities of the German Airborne
Corps. Although the combat narratives lack some tactical detail and no
substantial information on force structure is provided, the maps,
photographs, and technical information are excellent.

Glantz, David M., Lt. Col. The Soviet Airborne Experience. Leavenworth
     Paper no. 10. Fort Leavenworth, KS: Combat Studies Institute, U.S.
     Army Command and General Staff College, forthcoming.

     Once published, Lt. Col. Glantz's work will become one of the most
authoritative English-language sources available on Soviet airborne forces.
Using sources almost entirely Soviet in origin, with German sources for the
World War II operations, Glantz covers the time period from the 1920s to the
present day. This study is a valuable source of detailed information on
Soviet airborne operations in World War II at the tactical and operational
levels. Maps, tables, and data on airborne force structures are included.

Hickey, Michael. Out of the Sky: A History of Airborne Warfare. New
     York: Charles Scribner's Sons, 1979.

     This work goes beyond paratroop operations to consider the entire gamut
of airborne warfare, including troop delivery by parachutes, gliders,
helicopters, and reconnaissance craft. Although the focus is on airborne
operations conducted during World War II by all the major belligerents,
postwar developments are also discussed; the U.S. airmobile experience in
Vietnam receives special attention. The author's objective is to
demonstrate the utility, complexity, and changing nature of operations in
three dimensions.

Hoyt, Edwin Palmer. Airborne: The History of American Parachute Forces.
New York: Stein and Day, 1979.

Airborne is a cursory overview of the history of American airborne
forces from 1940 to 1978. Not really a scholarly work, the book is written
in a popular style interspersed with war stories and anecdotes. Although it
contains little analysis of the utility of airborne forces or airborne force
structures, it is useful as a general record of the activities of America's
own airborne units. Maps, photographs, and pictures of unit patches are
included.

Huston, James A. Out of the Blue: U.S. Army Airborne Operations in World
    War II. West Lafayette, IN: Purdue University Studies, Purdue
    Research Foundation, 1972.

Huston's work is acknowledged as one of the best studies of the
American airborne experience in World War II. He discusses the development
of American airborne formations in detail, as well as the training,
materiel, and doctrinal bases for the new arm. An operational history is
included. The author's analysis of the effectiveness of the airborne arm is
illuminating.

Luttwak, Edward N., et al. "Historical Analysis and Projection for Army
    2000." Pt. 1. Chevy Chase, MD: Edward N. Luttwak, 1982- .

This collection of individual papers has been cited previously. The
two papers described below concern Soviet and German airborne forces.

_____. Paper no. 9. "The Germany Army of the Second World War. The
    Parachute Troops: The Fallschirmjaeger Formations." 1 March 1983.

The distinguishing characteristic of German airborne troops, especially
relative to other airborne forces (U.S., British, Soviet), was their
development from a force used solely for raids into a true light infantry
noted for their tactical skill and agility. Organizational development,
operational methods, comparative uses, and the evolution of tactics are
discussed in this paper.

_____. "Soviet Airborne Troops, 1930-1983." 10 December 1982.

Soviet airborne operations have not been characterized by notable
success, yet the Soviet Army persists in maintaining a large number (seven)
of airborne divisions which are considerably lighter than the U.S. airborne
division. This paper reviews the Soviet use of airborne troops over the
past sixty years and describes the changes in organization, equipment, and
doctrine.

Whiting, Charles. Hunters From the Sky: The German Parachute Corps,
     1940-1945. New York; Stein and Day, 1974.

     A tribute to the daring, toughness, perseverance, and indomitable
spirit of German paratroopers. This book provides an operational history of
German airborne units in World War II. Well researched, it describes the
key figures in the development and employment of the German parachute corps
and illuminates some of the strategy and tactics of their use. Few maps are
included, and no information on force structure or weapon systems is
provided. The book serves reasonably well as a record of German airborne
operations but contains very little doctrinal analysis of the military art
embodied in the airborne capability.

Although many documents discuss Army force structures during the 1950s and 1960s, the items below are among the best and are the most germane to the current DA programs aimed at developing a new light infantry division force.

Cushman, John H., Maj. "The Pentomic Infantry Division in Combat."
     Military Review 37 (January 1958):19-30.

In a short survey of the capabilities and organization of the pentomic division, the author describes the improvements in mobility, combat power, and command and control that the pentomic division was supposed to provide. The article includes a discussion of offensive and defensive tactics for the new organization.

Perret-Gentil, J., Lt. Col. "Divisions--Three or Five Elements?" Military
     Review 41 (February 1961):16-25.

Perret-Gentil compares and contrasts the U.S. pentomic division with European and Soviet triangular divisions. In general, the author recommends the triangular organization because of its advantages in flexibility, simplicity, and command and control.

Scotter, W. N. R., Maj. "Streamlining the Infantry Division." Journal of
     the Royal United Services Institution 98 (November 1953):597-602.
     Digested in Military Review 34 (May 1954):89-94.

Maj. Scotter proposes that the best way to reduce the size of the infantry division is to organize platoons with five five-man sections, in lieu of three nine-man squad companies with four platoons, and battalions with four rifle companies and a support company. Brigade HQs should be dropped. Each division should fight seven infantry battalions.

U.S. Army Command and General Staff College. "Factors Determining Optimum
     Division Size (Implications of Small Divisions) CGSC 56-10." Fort
     Leavenworth, KS, 31 May 1956. CARL N-17981.13.

The small 8,600-man, pentomic-style division discussed in this short paper comprised up to five maneuver elements directly subordinate to the division headquarters. The advantages and disadvantages of fielding a small division during the nuclear age instead of a large division are discussed. The 1956 viewpoint on new technology versus tactical organization is interesting.

U.S. Army Command and General Staff College. "Optimum Organization of U.S. Army Divisions in 1960." Fort Leavenworth, KS, 27 February 1954. CARL N-17935.7.

In response to a directive by the Chief, Army Field Forces, the CGSC undertook this detailed study to investigate the optimum organization for Army divisions in 1960. It considers three types of divisions--armored, infantry, and airborne--and discusses the conventional and nuclear battlefield. It reviews the results of the post-World War II studies and conferences and considers weaponry, manpower ceilings, technological advances, and division missions and capabilities. This work is an important historical document in regard to the theoretical arguments on division organization in the 1950s.

U.S. Army Infantry School. "[Re]organization of the Infantry Division, 1960." Fort Benning, GA, 19 January 1954. CARL N-17935.5.

A companion piece to the CGSC study, "Organization of the Army During the Period 1960-1970," this document represents the proposal from the Infantry School for the optimum organization for the infantry division in 1960, incorporating the new equipment expected to be available at that time.

U.S. Army. 3d Infantry Division. "Final Evaluation Report on the ATFA Infantry Division (TOE 7T)." Ft. Benning, GA, 15 January 1956.

As the title indicates, the 3d Infantry Division tested the ATFA organization and reported its evaluation of the structures and capabilities of all echelons in the proposed division. Based on Exercise Sagebrush as well as other premaneuver training and testing, this study reports on the Army's first significant effort to deal concretely with infantry maneuver and logistics under nuclear conditions.

U.S. Army. 18th Infantry Regiment. 1st Battle Group. "Evaluation of the Capability of the ROCID Infantry Division to Fight Non-Atomic Wars." Report to the Commanding General, 1st Infantry Division. Fort Riley, KS, 21 May 1958. CARL N-17935.62-V.

As part of the implementation of the pentomic division, the 1st Infantry Division was required to test ROCID capabilities. The document above is a short report from the 1st Battle Group, 18th Infantry, to the CG, 1st ID, on a field test of the unit's ability to fight conventional combat. It provides some informative comments on command, control, and mobility of the unit and compares triangular and ROCID division capabilities.

Section V. COMPARATIVE VIEWS AND ALTERNATIVE PROPOSALS

Daluga, R. B., Lt. Col., et al. "The CONFAD Light Division." Paper by
    Work Group A, Section I, for elective course R460, "Development of
    Combat Divisions." Fort Leavenworth, KS: U.S. Army Command and
    General Staff College, Fall 1971. CARL N-13423.432-A.

A CGSC work group produced this study in 1971. It proposes a
10,000-man light division designed for operations in a non-European, light-
to mid-intensity scenario. The organization is austere, with three
mechanized and four leg infantry battalions and separate howitzer batteries
for each battalion.

Garfield, L., et al. Performance Effectiveness Comparison of the Air
    Assault Division with U.S. ROAD and Other Proposed Divisions. PRC
    R-674. Los Angeles, CA: Planning Research Corporation, 15 March
    1965. CARL N-18653.9.

Prepared under contract for the U.S. Army Combat Developments Command,
this study compares the air assault division with the ROAD airborne,
infantry, mechanized, and armored divisions. It also compares a proposed
(at that time) airmobile division and an infantry division supported by an
aviation brigade with the AAD. Both offensive and defensive missions were
analyzed. An apparent flaw in the study is the absence of significant air
defenses in the enemy force array.

Madison, Ervin E., Lt. Col., et al. "The Light Armored Infantry Division."
    Paper for elective course R460/0, "Development of Combat
    Divisions--Free World and Communist Powers." Fort Leavenworth, KS:
    U.S. Army Command and General Staff College, 1970-71. CARL N-13423.425.

The division proposed by this CGSC work group is designed to provide
more firepower over the same area than the ROAD division, but at a smaller
cost in manpower. It contains organizational diagrams, short descriptions
of changes from the ROAD structure, and justifications for those changes.

McGovern, Donald H., Col. "Realignment of the Pentomic Infantry Division."
    Student thesis, U.S. Army War College, Carlisle Barracks, PA, 19 March
    1959. CARL N-18627.14.

This AWC student thesis compares the pentomic (ROCID) division with the
triangular division of World War II and the Korean War. The study concludes
that both organizations are battle worthy and proposes that conventional
artillery, maneuver firepower, and communications be strengthened in the
pentomic division.

Shelton, Henry R., Maj. "The United States Infantry Division and the
    Australian Pentropic Division--Similarities and Differences." MMAS
    thesis, U.S. Army Command and General Staff College, Fort Leavenworth,
    KS, 18 May 1964. CARL N-19052.14.

    The U.S. short-lived MOMAR division is compared with the Australian
Pentropic Division. The U.S. division of 15,594 men was designed as a
general purpose force; the Australian division, patterned after the U.S.
pentomic structure, was designed for the Southwest Pacific and Southeast
Asia.

U.S. Army Intelligence and Security Command. Intelligence and Threat
    Analysis Center. Soviet and United States Division Comparison
    Handbook. IAG-35-U-78. Arlington, VA, 1 October 1977. CARL
    N-20151.19.

    About two-thirds diagrams, the INSCOM-prepared handbook compares U.S.
and Soviet infantry, armored, and airborne divisions down to company level
for tank/infantry forces and down to battalion level for combat support
forces. CSS elements are not compared.

U.S. Army Strike Command. "Infantry Division Tailored for Airlift."
    Fort Monroe, VA, 7 July 1962. CARL N-18888.2-B.

    This study proposes an infantry division specifically tailored for
strategic airlift into a secured airfield. Only 7,951 tons were permitted
in the design. As a result, the organization is very light, has reduced
mobility and no armor, and includes 1,215 vehicles with 867 trailers and
10,165 men. Also included is a discussion of a 4th Infantry Division
proposal for a light division.

Section VI. TECHNICAL ANALYSES

Costello, John, Maj. "The Strategic Implications of a High Technology Light
     Division as Part of the Rapid Deployment Joint Task Force." MMAS
     thesis, U.S. Army Command and General Staff College, Fort Leavenworth,
     KS, June 1982. DTIC and CARL ADB067736L.

     The author traces the development of the HTLD and discusses its
strategic implications as a component of the RDJTF. Because of the
technological force structure and its capability for strategic deployment,
it provides more conventional power along the escalation spectrum than
current forces.

Hamlin, F., et al. "Survey of Bridging Requirements for the Light
     Division." McLean, VA: BDM Corporation, 30 April 1982. DTIC and CARL
     AD-B064284L.

     The development of the High Technology Light Division (HTLD) as a
strategically mobile, state-of-the-art, high tech division required an
in-depth analysis of the bridging requirements in the theaters to which it
might be deployed. This study provides that analysis along with an
excellent review of the purposes, missions, and capabilities of the HTLD.

Lopez, Ramon. "The U.S. Army's Future Light Infantry Division: A Key
     Element of the RDF." International Defense Review 15, no. 2
     (1982):185-92.

     The author discusses the new equipment that could be included in a new
light infantry division. The 1,230 sorties of C5A and C141 aircraft
estimated as needed to move the HTLD could be reduced through the use of new
light armored vehicles (LAV) and other light trucks like those developed by
Austria, Germany, and Sweden.

Parsons, Larry D. "Airlift Support of the High Tech Light Division in the
     Contingency Area." Study project. Carlisle Barracks, PA: U.S. Army
     War College, June 1982. DTIC and CARL ADA118849.

     The HTLD final design calls for it to be deployable on 1,000 C141
sorties. Aspects of the contingency area and tactics to be employed by the
HTLD will increase reliance on tactical airlift for logistical support.
Increased use of LAPES and CDS is expected.

R. & D. Associates. "Light Infantry Mission Area Analysis." Draft RDA
     Proposal 579014. Marina Del Ray, CA, 29 Feb 1980. CARL N-19922.26-2.

     This work proposes a methodology for mission area analyses for light
infantry organizations. No conclusions are included in this draft proposal,
but most of the significant force design issues are identified.

15

U.S. Army Command and General Staff College. "Reduction of Vehicles in the Infantry Division." Fort Leavenworth, KS, 4 June 1953. CARL R-17860.1-A.

Charged with cutting 200 to 250 vehicles from the infantry division, the CGSC solicited comments from the infantry, artillery, armor, and transportation schools. The disagreements were numerous; a consensus was not reached.

U.S. Army Infantry School. "Draft Study on 'Combat Potential to Manpower Ratio.'" Fort Benning, GA, 9 November 1955. CARL N-18050.3.

Several different methods are offered for computing combat potential through an analysis of manpower ratios in Army organizations.

_____. "High Technology Test Bed [HTTB] Evaluation Plan for the Light Motorized Infantry Battalion." Fort Benning, GA, May 1983. DTIC and CARL ADBO73894L.

The LMIB was designed as a component of the HTLD capable of performing virtually any mission. The LMIB is armed with HMMWV/TOWs, MK-19s, Dragons, FAVs, and 4.2" mortars. As a 770+ man battalion, it does not appear to deserve to be called "light."

Section VII. FOREIGN ARMIES

Great Britain. Staff College. <u>Staff Officer's Handbook, 1984 Course</u>.
Camberley, Surrey: The Staff College, 15 July 1983.

Published for the British Staff College 1984 Course, this handbook
serves as a one-source reference for operational planning. It contains a
vast amount of unclassified data on equipment capabilities and
vulnerabilities, ammunition effects, unit organizations, and the like,
including information on the force structure of UK expeditionary forces.
The handbook is not widely available in the United States, being subject to
a limited distribution outside the Staff College.

Lucas, James Sidney. <u>Alpine Elite: German Mountain Troops of World
War II</u>. New York: Jane's, 1980.

Without doubt, this book is the best English language source on the
German mountain divisions of World War II. Based on extensive research and
interviews with some 200 former members of the Gebirgsjaeger, the book
provides both a stirring narrative of the exploits of these mountain units
and a wealth of information on their equipment, organization, and tactics.
It is an indispensable reference for persons interested in combat in
mountainous terrain.

Luttwak, Edward N., et al. "Historical Analysis and Projection for Army
2000." Pt. 1. Chevy Chase, MD: Edward N. Luttwak, 1982- .

This collection of individual papers produced under TRADOC contract
DABT-58-82-C-0055 has been cited previously. The papers described below
concern foreign nonstandard forces, mostly light formations, organized
during and after World War II.

_____. "Soviet Mountain Rifle Divisions of the Second World War." [Partial
Report]. 10 December 1982.

Between 1920 and 1960, the Soviet Army maintained a small number of
divisions specially organized, equipped, and trained for use in mountainous
terrain. The divisions were manned at levels from 9,500 to 12,500 and used
pack and man-portable heavy weapons and alpine personal equipment. The
rifle battalions numbered about 600 men each.

_____. "Soviet Motorised Anti-Tank Regiments and Brigades of the Second
World War." 10 December 1982.

One of the Soviet responses to the problems of defense against German
massed armor attacks was the formation of antitank regiments and brigades as
part of the mobile reserve of corps, army, and front commands. When
properly integrated operationally, these units were highly effective in
halting enemy attacks.

17

_____. "Soviet Airborne Troops, 1930-1983." 10 December 1982.

Soviet airborne operations have not been characterized by notable success, yet the Soviet Army persists in maintaining a large number (seven) of airborne divisions which are considerably lighter than the U.S. airborne division. This paper reviews the Soviet use of airborne troops over the past sixty years and describes the changes in organization, equipment, and doctrine.

_____. Paper no. 8. "The German Army of the Second World War. The Mountain Troops: The Gebirgsjaeger Formations." 1 March 1983.

Considered by the author to be the most versatile formations of the Wehrmacht, the mountain divisions performed extremely well in all kinds of close terrain. Many features of the mountain division--innovative tactics, exploitation of agility, avoidance of enemy firepower--are relevant today as possible models for light infantry formations and particularly for RDF forces.

_____. Paper no. 9. "The Germany Army of the Second World War. The Parachute Troops: The Fallschirmjaeger Formations." 1 March 1983.

The distinguishing characteristic of German airborne troops, especially relative to other airborne forces (U.S., British, Soviet), was their development from a force used solely for raids into a true light infantry noted for their tactical skill and agility. Organizational development, operational methods, comparative uses, and the evolution of tactics are discussed in this paper.

_____. Paper no. 10. "Urban-Warfare Task Forces (Kampfgruppen) and Emergency Ad Hoc Forces (Alarmeinheiten)." 1 March 1983.

To meet both planned and unplanned requirements for close combat in urban areas, the German Army in World War II formed ad hoc and formally structured urban-warfare units. Many of these units were regimental size and smaller, although a number were division size or greater (Stalingrad, Sebastopol). The organizational history and tactics of these units are analyzed in this paper.

The following papers, nos. 11-18, concern contemporary light infantry and mountain forces.

_____. Paper no. 11. "The Swedish Norrland Brigades and Jagar Units." 1 March 1983.

The Swedish Norrland brigades have been especially organized to fight in the sub-Arctic, varied terrain of the Swedish north. The brigades are characterized by their heavy use of all-terrain vehicles and use of local recruits. The primary unit of the brigades is the reduced 566-man Jagar battalion.

_____. Paper no. 12. "The Swiss Mountain Divisions." 1 March 1983.

The Swiss perspective on mountain warfare is reflected in the organization of the three Swiss mountain divisions. This paper maintains that the Swiss tactical emphasis is on fixed defenses, positional forces (versus agile, mobile, foot forces), and concentrations.

_____. Paper no. 13. "The Austrian Mountain Battalions and the Jagdkampf Forces." 1 March 1983.

Austrian plans for the use of ultra-light infantry in a frontless defense against invading forces is unique among modern armies. The concept and plans for implementation are deemed unrealistic, although the organization of the light infantry has some enlightening aspects.

_____. Paper no. 14. "The West German Light Infantry Forces Since 1956." 1 March 1983.

The German Army Structure 4 provides for one mountain infantry brigade and three heliborne infantry brigades. This study covers the development, organization, and intended uses of these light infantry forces since 1956.

_____. Paper no. 15. "An Exposition and Critique of the West German Mountain Warfare Manual (Die Kompanien des Gebirgs-Jaegerbat-taillons)." 1 March 1983.

Current German offensive and defensive mountain tactics are analyzed in this short article.

_____. Paper no. 16. "Notes on the Israeli 35th (Paratroop) Brigade and Derived Reserve Brigades, with Additional Notes on the 'Air-Landed Force' and the Golani Brigade." 1 March 1983.

Israeli paratrooper forces are intended for use primarily for operations in close terrain at night or by helicopter insertion, as opposed to long-range airborne or air-landed operations. The combat records of the 35th (Paratrooper) Brigade and derived reserve brigades are examined in this paper.

_____. Paper no. 18. "Notes on Special-Purpose Forces, Dissimilar Formations and Expeditionary Headquarters in the British Army." 1 March 1983.

Because of its historical involvement in small-scale, extra-continental forays throughout the "Empire," the British Army has developed "dissimilar" terrain-specialized and task-specialized forces in contrast to general purpose forces. Their experience in the use of these forces represents a large store of valuable knowledge.

_____. "Part Two: Analysis and Conclusions." (Draft final report).
15 March 1983.

The final report in the Luttwak series of papers contains extensive historical analysis drawn from the individual research papers above. The author draws conclusions regarding the most advantageous uses and organizations for light infantry in modern warfare.

Luttwak, Edward, and Dan Horowitz. The Israeli Army. New York: Harper & Row, 1975.

The military history of the Israeli Army from its inception in 1948 through its participation in the October War of 1973 is covered in this book. Its utility to persons interested in light forces is found in its limited discussions of the development, training, equipment, force structure, tactics, and operations of the airborne units and the Golani Brigade.

Middleton, James, Lt. Col., et al. "The Battle for Darwin/Goose Green." Staff group battle analysis. Fort Leavenworth, KS: U.S. Army Command and General Staff College, May 1983. Combat Studies Institute files.

This unpublished CGSC staff group battle analysis is an in-depth look at the battle for Darwin and Goose Green in the recent Falklands War. The story of the deployment, commitment to battle, and support of the British light forces that won this battle contains some timely lessons on low- to mid-intensity, long-range intervention worthy of close study. The paper includes a three-page bibliography on the Falklands War.

Rothenberg, Gunther E. The Anatomy of the Israeli Army: The Israel Defense Force, 1948-78. New York: Hippocrene Books, 1979.

The author describes the historical background of the Israeli Army and records the history of its formation, organization, and operations from 1947 to 1978. Of particular interest to the reader are those passages dealing with the airborne forces and the many dismounted combat actions. However, the author does not present in-depth analyses of force structures or of small unit actions.

U.S. Defense Intelligence Agency. Handbook on the Chinese Armed Forces. Washington, DC, July 1976. CARL N-18911.557.

The large majority of the Chinese forces are dismounted infantry organized along Soviet lines. The absence of mechanized and armored formations is due more to economic and industrial factors than to a disposition for dismounted operations, although large areas of China are favorable for light infantry.

D 119953

www.ingramcontent.com/pod-product-compliance
Lightning Source LLC
Chambersburg PA
CBHW081342090426
42737CB00017B/3257

* 9 7 8 1 6 0 8 8 8 0 7 0 6 *